George Nelson Godwin

The Geology, Botany, And Natural History of the Maltese

Islands

George Nelson Godwin

The Geology, Botany, And Natural History of the Maltese Islands

ISBN/EAN: 9783744717625

Printed in Europe, USA, Canada, Australia, Japan

Cover: Foto ©berggeist007 / pixelio.de

More available books at **www.hansebooks.com**

Ex-libris

Charles Atwood
Kofoid

THE

GEOLOGY, BOTANY,

AND NATURAL HISTORY

OF

𝕿𝖍𝖊 𝕸𝖆𝖑𝖙𝖊𝖘𝖊 𝕴𝖘𝖑𝖆𝖓𝖉𝖘.

COMPILED BY

The Rev. G. N. Godwin,

CHAPLAIN TO THE FORCES.

MALTA:

PRINTED BY PAOLO BONAVIA:—M.DCCC.LXXX.

SOLD BY W. WATSON, 248 Strada Reale.

TO

CHARLES AND EMILY BURT,

MY FRIENDS AND TEACHERS,

THIS LITTLE WORK

IS DEDICATED.

CHAPTER I.

GEOLOGICAL OUTLINE.

Ancient and Modern Geologists.—Classification of Strata. Upper or Coral Limestone. — Yellow Sandstone and Blue Clay.—Free-stone.—Lower Limestone.—The Great and Lesser Faults.—Fossil Fauna.

THE Knights of St. John had in their ranks some few observers of natural phenomena. In 1647 the historian Abela wrote quaintly concerning certain huge bones which he somewhat credulously supposed to be those of the giant builders of the ancient temples of Malta and Gozo, and in 1747, Scilla, the well known Sicilian artist, pourtrayed several characteristic Malttse fossils and the teeth of the carnivorous fossil whale Zeuglodon, the remains of which have also been met with in North America. The Commander Dolomieu about a century ago laboured in the same field, but it is to the patient industry of Professor Edward Forbes, Captain Spratt R. N., Dr. A. L. Adams M. A., Dr. Wright and others, that we are chiefly indebted for our knowledge of the various geological formations of the Maltese islands, and it is from their works

that the following outline has been compiled. Herr
Fuchs must also be mentioned as the author of a
very able sketch on "The age of the Tertiary Formations in Malta, *(Das alter der Tertiären Schich-
ten für Malta)*, which contains the most complete
list at present in existence of the various fossils
found in the different strata.

The Maltese islands situated nearly in the centre
of the Mediterranean basin, are of Tertiary origin
and are now generally admitted to belong to the
late Eocene subdivision of that formation. It is
scarcely necessary to remark that the names of
Eocene and Miocene were given by Sir Charles
Lyell to the Lower and Middle Tertiary strata. The
Eocene fossils include only 3½ per cent of living
species whilst in the Miocene formations the proportion of living to fossil species is 25 per cent.
Dr. Adams says that "the Maltese islands are as-
suredly mere fragments of what had once been an
extensive sea-bottom, which when first upheaved
formed part of either Europe or Africa or both,
and lastly that after oscillations of level the greater
portion was submerged, leaving only these small
remnants now known to us as Malta, Gozo, and
Comino." The latest theories favour the idea of upheaval after previous submergence. To quote
Captain Spratt R. N. "The axis of the chain
of the Maltese islands runs from S. E. to N.
W. and is about 29 miles long. Malta, the south-

ernmost of the group, is nearly 17 miles long; and its greatest breadth measured transversely to the axis, is nearly 9 miles. Gozo, the northernmost island is nearly nine miles long, and its greatest transverse breadth is a little more than 5 miles."

"The mineral deposits of which these islands consist are all stratified and disposed in parallel layers. They seldom deviate much from the horizontal position, but the prevailing dip, which is very gentle, varies from N. E. to E. by N. and consequently the prevailing strike of the deposits coincides nearly in direction with the axis of the chain."

The Apennines and the Sicilian chains have the same inclination rendering it probable that all were upheaved at the same time. The rocks blend into one another so gradually that it is sometimes not easy to say where limestone ends and sandstone commences.

"None of the deposits are wholly devoid of organic remains and some of the softer strata contain them in great abundance, and in a state of excellent preservation. Many of these fossils are characteristic of certain strata, and all are of marine origin."

From the discovery of the remains of amphibious animals such as the dugong, manatee, seals, and crocodilians, together with corals and corallines it is thought that the rocks of Malta were formed at no great distance from land.

Captain Spratt thus groups the strata in descending order: 1 Coral Limestone: 2 Yellow Sandstone and Blue Clay: 3 Freestone: 4 Semi-crystalline Limestone: whilst Dr. Adams has five divisions thus: 1 Upper Limestone: 2 Sand: 3 Marl: 4 Calcareous Sandstone: 5 Lower Limestone.

The Upper or Coral Limestone (A) which is in some places 250 feet in depth "consists of a reddish-brown or whitish calcareous rock which is mostly of a compact, hard and almost flinty texture. It contains cretaceous nodules, and is sometimes interstratified with soft calcareous sandstone." It covers almost the entire S. W. and N. W. portions of Malta. The island of Comino with its cliffs rising more than 200 feet above sea-level is almost entirely composed of it, and it forms a capping to nearly all the isolated hills of Gozo, from which however t has been much denuded. "Near Casal Garbo, towards the N. W. angle of Gozo, the only remains of the coral limestone, which originally formed a continuous upper crust over that part of the island, are detached masses of this deposit lying on the surface of denuded freestone. Similar masses are seen in other parts of the island. Some of these fragments at Casal Garbo are variegated with yellow and white, and are used for ornamental work under the name of "Gozo Marble."" It abounds in fossils, amongst which molluscs and echinæ are numerous with some remains of sea-weeds, but traces of the higher animals

THE GEOLOGICAL FORMATIONS
OF THE MALTESE ISLANDS.

No. I.	A...	...Coral Limestone.
	B...	...Yellow and black or green sand intermixed.
No. II.	C...	...Marl.
	D...	...White sandstone.
	E...	...Reddish, yellow, & grey sandstone.
No. III.	F...	...Pale yellow sandstone.
	G...	...Chocolate-coloured nodules, teeth, shells, &c.
	H...	...Yellow sandstone.
No. IV.	I...	...Semi-crystalline limestone.

The geologist should not fail to examine the admirable Geological Map at the Garrison Library, and the Geological Sections in the Museum, Valletta.

are not so frequently met with as in the under-
lying beds.

The Yellow Sandstone (B) the depth of which
is variously estimated at from 20 to 40 feet "consists
of yellow sand or sandstone with greenish-black
particles intermixed. It abounds in organic remains
many of which differ from those of the coral limestone.
One of its most characteristic fossils is a small
Nummulite which sometimes is in such quantity
as to form a third part of the bed to which it
belongs. It occurs most abundantly in the cliffs
of the Bay of Ramla on the N. W. coast of Gozo."
Layers of oysters, the teeth and bones of sharks,
and the remains of Cetacea are also met with in
this formation which can be easily examined in the
cliffs on the N. W. shore of Malta, at Ramla Bay in
Gozo, and below Fort Chambray in the same
island, and also in the neighbourhood of Città Vecchia.

The Blue Clay or Marl (C) from 100 to 120
feet in thickness "contains two or three thick layers
of a lighter colour than the rest, and imbedded
in it are crystals of gypsum and occasionally nodules of
sulphur." The cliffs at Karabba shew to advantage
the thickness of the marl, which is strougly impregnat-
ed with lime, and contain a few organic remains.
The bones of whales, sharks' teeth, teeth and spines
of rays and other species of fishes are not uncommon.
"The Testacea are chiefly species of Mitra etc.
A Nautilus is found, but rarely, under Fort Chambray.

With the shells has been found the bone of a small Sepia (or cuttle fish). The fossils of the clay generally serve as nuclei to irregular nodules of iron pyrites, and the substance of the fossils is frequently converted into hydrated peroxide of iron."

To the Blue Clay succeed five beds which Captain Spratt groups under the name of Freestone and Dr. Adams under that of Calcareous Sandstone, with a depth of 200 feet. This formation occupies fully one half of the surface of both islands, and in Malta its district would be portioned off by a line running about N. through Città Vecchia.

"The clay passes into a white calcareous sandstone (D) from 20 to 30 feet thick, and below this is a blueish-grey or fawn coloured marl (E) about 20 feet thick. These two deposits contain several species of microscopic chambered shells."

"Next are found from 20 to 30 feet of a pale yellow or white calcareous freestone (F) separable into thin strata. It contains nodules of flint, and the fossils of this bed are found in a silicified state on the N. W. side of the Bengemma hills."

"This stone is sometimes used for building, but it exfoliates by exposure to the weather, and more particularly when acted on by the sea. It contains a Scalaria and other forms."

"Below the upper bed of Freestone is a bed, from 2 to 8 feet thick of Calcareous Sandstone (G) of a pale chocolate colour and flinty hardness which

consists almost wholly of the casts of organic remains (see Professor Forbes' catalogue), and mixed
with the casts are shapeless nodules of the sandstone of the same colour and texture. This deposit
preserves its peculiar character wherever the freestone group of beds is found. It is best exhibited
in the island of Gozo, in the Bay of Marsa el
Forno on the N. W. coast, and at the base of the
cliffs under Fort Chambray where it forms rocky
ledges two or three hundred yards broad, extending
along the coast, and rising only a foot or two above
the sea level."

"The lowest bed of the group is a yellowish
white calcareous freestone (H) from 40 to 50 feet
thick. This is the stone which is commonly used
for building in the two islands. From the facility
with which it may be cut with the hatchet, or formed ·
with the lathe, this stone both in the rough state
in the form of slabs, and also when turned into
pillars, balustrades, vases, and other architectural
ornaments is used extensively in all the public and
private edifices of Malta and Gozo, and is an arti-
cle of considerable export to all parts of the Mediterranean. A fossil turtle was found in this bed
near Casal Luca, south of the city of Valletta."

Remains of the great carnivorous whale Zeuglodon, of seals, and of the amphibious mammal
Halitheium have been discovered in this formation,
together with jaws of crocodilians, not to mention

numerous molluscs, cuttle fish, barnacles, and marine plants.

The lower or semi-crystalline limestone (I), is of a greyish colour, and on the S. and N. W. coasts of Gozo shews nearly 400 feet of perpendicular depth. It, in common with the harder varieties of the Upper Limestone is known as Gozo marble and Malta granite. It is extensively quarried for building purposes in the neighbourhood of Musta, and on the denuded flat to the W. of Valletta.

All along the southern shores of tho islands, this formation is gradually yielding to the disintegrating influence of the Sirocco blast, and the ever-beating surf.

The Lower Limestone can be easily studied between Fort Ricasoli and the Zoncor Tower, whilst inland it may be traced in the Wied Incita, near the Lunatic Asylum, in the ueighbourhood of Musta, and below Gargur.

Fossils are abundant, but owing to the hardness of the rock perfect specimens are detached with difficulty. The saucer shaped urchin, an organism resembliug fossil leaves, sharks' teeth, whales' bones, oyster shells, claws of swimming crabs, burrowing sponges, &c., are amongst those most frequently met with. Nine species of shark formerly inhabited these waters, the teeth of some of them being 7 inches long!

Malta is divided into two parts by a great "fault" or break in the continuity of the strata, due to either depression or upheaval. This fault "cuts the island transversely to the axis of the chain, and to the N. W. lets down the strata about 300 feet. Gozo also is divided, a little way inland from the strait which separates it from Malta by a fault running also transversely to the axis of the chain, and producing to the S. E. nearly the same amount of depression in the strata which is occasioned in the opposite direction by the fault of Malta. The joint effect of these two disturbances is to let down the deposits in the space between the two faults to the depth above-mentioned, that being about half the height above the sea-level of the most elevated points in each of the two islands."

A rising ground near Casal Dingli 750 feet above sea-level, and the hill of Bisbiegi with an altitude of 743 feet are the two highest points in Malta and Gozo. "In the sunken tract lie the straits of Frieghi which separate Malta from Gozo, and midway between the two principal islands, the small island of Comino." The fault of Malta is clearly visible in the Bay of Fom-e-Rieh (mouth of the wind) on the S. W. shore of the island, from whence it passess below the Bengemma Hills, crosses the plain of Nasciar, and reaching the northern shore at Maddalena terminates in a bold bluff. (See Section 4.)

The Revd. H. Seddall says: "To the existence of this fault is due one of the most picturesque features of Malta. Often on a fine spring morning have I stood on the ridge of the Nasciar heights, the whole plain below glowing with the blossom of the purple sulla (a kind of clover *Hedysarium coronarium*), the inlets of the bays of the Salini, St. Paul's, and the Straits of Frieghi reposing like gems of the deepest blue in their setting of white rock, which the sun irradiated with a perfectly dazzling lustre, enjoying the first cool breath of the *maestrale*, as it dimpled the azure of the lazy deep, and mapping out the course of present or future excursions with gun, hammer, or botanical box."

Minor faults in the depressed area have aided in the formation of the Bays of St. Paul and Melleha. According to Capt. Spratt the Gozo fault emerging from the ravine of Ghain Selim passes to the N. of Fort Chambray, and to the S. of Casal Nadur, meeting the sea on the N. E. coast in the Bay of Silek (See Map and Section 2), but Dr. Adams assigns to it a somewhat different direction. Numerous minor faults exist such as that of Malak on the southern shore, the Macluba near Crendi, and the curious hollow known as the Kaura in Gozo.

"Most of the valleys in the Maltese islands

follow the course of the dip of the strata. Among the
Bengemma Hills lie the valleys of Boschetto and
Emtahleb, which are noted for their picturesque
scenery and also for their fertility. Their productiveness
is owing to the springs which break out at the
outcrop of the blue clay, in consequence of its
retaining the moisture which falls on the porous
substance of the superincumbent coral limestone and
yellow sandstone. There are no springs in Malta
and Gozo but where there is clay to retain the
water." The washing away of the marl would
therefore make Malta a mere arid rock and deprive
Valletta of its supply of water.

From the numerous and interesting remains of
hippopotami, elephants, fresh-water tortoises etc.
found in these islands, and also from other evidence
it seems probable that Africa, Malta, and Italy were
formerly united and that the now submerged district was covered with a varied and extensive
flora. Herds of both larger and pigmy elephants the
latter being 3 feet, 4 feet 7 inches, and 7 feet in
height roamed the land. Dr. Adams and Sig. Caruana
have found skeletons or teeth of hundreds of these animals near Benghisa Tower, near Fort Ricasoli, at Zebbug, Crendi, the Mnaidra Gap, the Malak and Melleha
caves, and other places.

The *Hippopotamus Pentlandi* has left his bones
at Melleha, Malak, and elsewhere. Abundant wild-fowl
covered the lakes and lagoons, one of which the

Cygnus Falconeri was fully one third larger than our present swan. Its remains have been found at Gandia and Mnaidra. The *Myoxus* a gigantic fossil dormouse, as big in proportion to the living species as a good-sized rabbit is to a brown rat existed in large numbers. A lizard larger than a chameleon and a huge fresh water tortoise have also been met with. Carnivorous animals were not wanting for the elephants' bones found at Zebbug showed marks of fierce and eager gnawing. The bones of a ruminant allied to the goat or sheep have been unearthed near Crendi. No traces of implements or of man's presence amongst these long extinct animals have as yet been discovered. Those who wish to pursue this subject further are referred to the " Notes of a Naturalist in the Nile Valley and Malta" by Dr. Adams, Captain Spratt's " Geology of the Maltese Islands, " Seddall's " Malta Past and Present, " Tallack's " Malta under the Phœnicians, Knights, and English" and other kindred works too numerous to name.

CHAPTER II.

Botanical Notes.

Maltese Botanists.—Books of Reference.—General Aspect of the Islands.—Trees and Fruits.—Marine Flora.—Grasses and Aromatic Plants.—Midsummer in Malta. — Spring and Summer Vegetation. — Fungus Melitensis, etc. — Botanical Excursions in Malta.—The Flora of Gozo.

IN the year 1670, Dr. G. F. Bonamico the Father of Maltese Naturalists enumerated 243 distinct varieties of plants as existing in these islands. Five years afterwards Dr. G. Zammit a member of the Order of St. John occupied the Botanical Chair, and established a Botanical Garden in the moat of Fort St. Elmo. Dr. Cavallini, one of Dr. Zammit's pupils published in 1698 the result of the researches of Bonamico, with the addition of 83 varieties noted by himself. P. Boccone of Palermo between 1674-1697 wrote ably upon the flora and fauna of Malta. He has been styled the Pliny of his century. Not to mention others, Peter Forskal in 1775 described the plants of Egypt and Malta, and was followed by J. D' Urville in 1822. In 1825 G. E. Giacinto

a Genoese, who had been appointed Professor of
Botany in 1805 by Sir Alex. Ball, published with
the aid of Drs. Naudi and Zerapha "The Plants of
the Islands of Malta, Gozo, and Lampedusa" enu-
merating 854 distinct varieties.

In 1827 and 1831 Dr. S. Zerapha published
his admirable "Floræ Melitensis Thesaurus" which
describes 498 native and 155 exotic plants, and in
1853 the "Flora Melitensis" of Dr. Delicata containing
an account of 716 plants saw the light. Three
years later Dr. Gulia published his valuable "Repertorio Botanico" to which the student is referred,
and from which many of the foregoing particulars
have been gathered. See also Dr. Gulia's "Repertorio
di Storia Naturale," Gussone's "Floræ Siculæ Syn-
opsis" (Naples), and Wood's "Tourists' Flora" (Reeves,
Henrietta Street, Covent Garden), together with
"Malta Past and Present," by the Revd. H. Seddall, and Dr. L. Adams' "Notes of a Naturalist
in the Nile Valley and Malta," and the exhaustive
Analytical Flora of Dr. Gulia, one third of which
has already (1880) seen the light.

The islands have a general aspect of barrenness,
owing to the want of trees, and the absence of
shade renders the heats of summer more than
usually oppressive. The three necessaries for successful arboriculture, viz., abundant soil, constant
moisture, and shelter, are not always readily attain-
able. But where these advantages exist, almost all

trees of temperate or even tropical regions flourish and thrive. Beneath the surface rocks there is an inexhaustible bed of marl eminently adapted for the support of trees. The Gozitans were formerly in the habit of destroying all trees along the roads lest their spreading roots should lessen the yield of cotton from the adjacent fields. The carouba is the most common tree, but it abounds chiefly in the eastern districts of Malta. The aloe, cactus indicus, and geranium attain a considerable size. The principal fruits are strawberries, figs, pomegranates, grapes, prickly pears, apples, pears, peaches, nectarines, apricots, plums, cherries, melons, and lemons. Also the Japanese medlar or nespoli and the orange, which latter is not surpassed anywhere in the Mediterranean. Even the best varieties of grapes speedily degenerate in these islands.

The planting of trees would lessen the excessive heat, increase the water supply, and multiply fruits, vegetables, and flowers. The indigenous plants of which Dr. Zerapha enumerates 644 species are more numerous than might perhaps be expected from the rocky nature of the soil and the almost universal cultivation. The marine flora are of course numerous. The grass wrack is washed on shore in vast quantities, and does good service as manure. Those however which require a sandy beach are comparatively rare such as *Polygonum maritimum* found at the Marsa and St. George's Bay, *Cakile maritima* in the Bay of Melleha, *Euphorbia Para-lias* in the Bays of Saline and Gneyna, *E. terra-cina* in Melleha Bay, *Eryngium maritimum* at Mel-leha, Gneyna, etc., *Pancratium Illyricum* at Ramla Bay in Gozo. Dr. Zerapha enumerates 19 species of the handsome *Euphorbia*. A very common marine plant which loves the sandy shore is the *Crucia-nella maritima*, which may be met with at Sliema and elsewhere. It blossoms in April and June. On the rocks of the southern shore are particularly to be noticed the *Hypericum Ægyptiacum* and *An-thyllis Hermanniœ*, the latter having also an affect-ion for the barren hill side.

The islands are rich in the natural order of the *Papilionuceœ*. Of this the genus *Trifolium* counts the greatest number of varieties of which some of the most interesting are *Trifolium subterraneum* found in sunny spots during March and May, and *Trifolium suffocatum* not unknown at Floriana. Then follow the genera *Medicago, Melilotus, Lotus, and Ononis*.

The grasses are of course in great variety. We may note the *Lygeum spartum* found at St. Paul's Bay, Imtahleb, Fauara, etc. The *Stipa tortilis* and *Stipa pinnata* are found everywhere on uncul-tivated spots.

Aromatic plants are few in number but we may note *Mentha Pulegium vulgare, Melissa officinalis, Nepetha calamintha,* and *Thymus capitatus*, the flowers of which give a delicious flavour to the honey of Malta whilst the stalks are used as fuel.

Dr. Adams says. "The physical aspect of the Maltese islands in midsummer is by no means inviting. Viewed from a commanding position, they present an extremely sterile and desolate appearance which is heightened by the interminable stone walls, rocky ravines, bare plateaus, and plains without marsh or stream; for, excepting a few fig, vine, cactus, carob, orange, pomegranate, and Persian lilac trees in gardens in or about the towns and villages not a blade of grass or a plant of any sort is there to gladden the eye, or relieve the glare of a semi-tropical sun."

But after the autumn rains Malta grows green as if by miracle. In January, anemones, several varieties of *Fumaria*, geraniums, the *Hypericum Ægyptiacum* and numerous other plants are in blossom. The Mediterranean heath *Erica peduncularis* is found in the Wied Incita, and the borage, rosemary, various euphorbias, and plants of the nettle tribe will repay the toil of the botanist, not to mention the narcissus and asphodel.

In February the pheasant's eye, poppies, mallows, and geraniums, vetches, chrysanthemums and varieties of the iris are in bloom.

To quote Dr. Adams once more "As far as verdure is concerned Malta may be said to be in its prime in February. It is then that the daisy

and dandelion deck the meads, grassy lanes, and
waysides, wheat is ripening, and the luxuriant tops
of the purple vetch *Hedysarium coronarium* adorn
the terraced fields and commingle their flowers with
the red poppy, yellow marigold, daffodil, crimson
pheasant's eye and purple anemone, where the painted
lady, cabbage, clouded saffron, and other butterflies
are sporting. The evergreen of the stunted locust
or carob tree (*Ceratonia siliqua*) contrasts well with
the whole scene, whilst the bare boughs of the
fig stand out in inanimate ugliness against the
stone fence around the terraced fields. About the
15th of February wheat is in ear, and the progress in
vegetation may be said to have reached its height."

During March and April vegetation is luxuriant.
Ranunculi, poppies, several species of *Cruciferæ*, etc.
flower. The caper plant makes fortifications gay
with leaves and flowers, and the ice plant may
be met with.

The summer flowering plants mostly belong to the
Compositæ. The so called *Centaurea Spatulata* with
leaves like those of a *Sempervivum* which grows on
the rocks above Fauara, blossoms in May. Some
botanists deny its claim to be considered a *Centaurea*
and regard it as a distinct genus. Snapdragons,
aromatic plants, glassworts, euphorbias, the caper,
cinerarias, etc., with numerous others perfume the air.

Fifteen species of the orchid tribe amongst which
are the *Orchis undulatifolia*, and *saccata* are found.

The *Ophrys tenthredinifera* or saw fly orchis is re-markable; gladiolus, wild garlic, squill, 17 species of sedges and 77 grasses are also to be procured. The viscous *Orsinia camphorata* with a strong smell of camphor grows on the walls of Valletta, and in many a valley.

Other rare plants are the *Scolopendrium hemi-onitis, Fagonica cretica, Putoria calabrica* found on a rock near the chapel of St. Paul the Hermit in the Wied El Ghasel, near Musta, *Convolvulus cala-brica* which grows near Imtahleb, *Cheiranthus tricus-pidatus* near Marsa Sirocco, *Teucrum scordioides, Helianthemum Fumana* near Gerzuma, *Hyacinthus romanus* at Fauara, Imtahleb, and Musta, *Carthamus cœruleus* at Imtahleb, etc.

The *Cynomorium coccineum* or *Fungus Meliten-sis* which is found on the southern shore of Malta near Casal Dingli, and more abundantly on and around the General's Rock in Gozo, was long in great repute as a remedy for hœmorrhage and dys-entery. It begins to blossom in April. It is said that the *Daucus gummifer,* certain species of the *Cheir-anthi* and *Gnaphalia,* and several other plants not met with elsewhere in the Maltese islands are to be found on the General's Rock.

During the summer months botanists who care not to suffer from sunstroke will not go far from home. Dr. Delicata's "Flora Melitensis" admirably points out the localities in which various plants will

be found, together with their times of flowering,
and Dr. Gulia in his Analytical Flora estimates the
Maltese plants to be not less than 1000 in number.

The following remarks by the last named able
naturalist will be found useful by the botanist in
his rural excursions.

" Let us ascend into the higher valleys which
formerly when submerged gave shelter to marine
animals. Here we shall meet with the fragrant
narcissus and the modest violet, which take the place
of the seaweeds which formerly covered this district."

"Let us penetrate into Wied Kirda, on the stately
sides of which the *Coronilla valentina* blossoms in
March. Its perfume which resembles that of the jon-
quil is a good reason for its cultivation within doors.
Some authors say that it exhales perfume by
night, but it is certain that it is also fragrant
during the day. Very beautiful is the yellow tulip
Tulipa sylvestris which grows beneath the olive, and
on the edges of the fields opposite to the picturesque
Chapel of Sant. Antonio. It was first met with
here by Dr. Agostino Naudi, and grows nowhere
else in Malta or Gozo."

" The ivyclad walls of the fields urge the botanist
to continue his researches, which will be repaid by
Calendula sicula with its large flowers and by
many other species. One of the best places for a
botanical excursion is the Wied Babu, between
Crendi and Zurrico, where, at various seasons, a

large portion of the Maltese flora can be studied. Large bushes of the fragrant rosemary cover the rocks, and here and there upon heaps of stones may be seen the *Lonicera implexa* with its twig-abounding, climbing stalks adorned with large masses of blossom. The beautiful *Gypsocallis multiflora* is abundant and most beautiful orchids, two kinds of narcissus, and several coronillas make gay the soil. The service tree, the quince, the white thorn with its numerous varieties, the Neapolitan medlar, together with the cultivated and wild pear tree grow there, no doubt spontaneously. We must not forget to mention a very beautiful *Centaurea* which grows on the rocks near the sea, which from the form and thickness of the leaves has been variously named *spatulata* and *crassifolia*. From the beauty of its large violet-coloured flowers it deserves to be introduced into our gardens."

"Amongst rare plants we may mention the *Scorzonera octangularis* found at Uardia by Mr. C. A. Wright, who also discovered the *Ononis variegata*, and in Gozo the *Senencio*.

The *Aristolochia longa* is found in the Wied Herief, and the *Scolopendrium officinale* in the Wied Ghomor. On a shady rock in the Wied Babu the *Splenium trichomanes* finds a home, and the *Romulea ramiflora*, *Asparagus fistulosus*, *Iris tuberosa*, and *Sisymbrium sophia* are not unknown." With reference to the flora of Gozo Dr. Gulia says :—

"After making several excursions in various directions the botanist will observe that in districts in which the upper calcareous strata are found, plants thrive which are sought for in vain in the denuded districts: and that in the localities where marl is abundant he will find plants which do not grow elsewhere. It is on this account that more wild varieties are met with in Gozo than in Malta, and that others which in Malta are sickly and captious, in Gozo exhibit all the vigour of healthy growth, and cover a considerable area."

"The vegetation of the N. W. portion of Malta resembles that of Gozo, the *Kundmannia sicula* which is found near Città Vecchia and at St. Paul's Bay, growing also in great abundance upon all the calcareous hills in the sister island. The beautiful *Reseda lutea* which in Malta is found in the spots loved by the *Kundmannia* is very plentiful in Gozo. Numerous also are the plants which grow in Gozo and not in Malta. The sweet and vigorous balm mint *Melissa officinalis* clothes all the valleys, and there grows also in the clefts of the rocks of the Wied el Xlendi the *Silene fruticosa* with its rosy flowers, and which at first sight seems to be a saponaria, together with *Raphanus landra* and *Raphanus fugax* by the side of the never-failing brooklet."

"In the valleys wherever running water is to be found the agnus castus or *Vitex agnus castus* flowers in profusion. Its seeds were formerly considered to be of use in checking immodesty, and for the preservation of chastity, which caused the monks of the fifth century to cultivate it in their monasteries with jealous care. It grows vigorously at Marsa Scala and Wied el Buruni together with the *Populus alba* and the *Tamarix Africana*. On several hills in Gozo thrives the graceful little valerian named by Dufresnoy *Centranthus Calcitrapa*."

"Unless the botanist visits Gozo in the springtime he can form no idea of the beauty and wealth of some of its valleys in which the fertilising waters constantly refresh numerous indigenous plants, which together with the cultivated species make these districts fair to look upon. The *Hypericum Ægyptiacum* grows abundantly all adown the valley of Xlendi, the steep sides of which here and there give sustenance to the *Daucus australis*. In the same valley are to be found very beautiful specimens of *Delphinium longipes, Chlora perfoliata, Sedum amplexicaule,* and *Sedum cœruleum* which the Gozitans call, I know not for what whim, Gheneb il Madonna or Our Lady's grape, the rare *Scutellaria peregrina*, and many another plant which adds to the beauty which Nature has lavished upon this spot."

"Plants belonging to our flora grow abundantly in like manner in other valleys. In marly soils, and in the neighbourhood of the sea may be recognised by its rosy red cluster of blossoms the lesser centaurea, which is no less useful in medicine than lovely in the garden. At El Pergla in the proper season we may find the *Triolium abbreviatum*, the *Scabiosa longiflora*, the cineraria with its golden clusters, and the *Samolus Valerandi*. There the queen of European grasses assumes stately proportions, and there grow in great abundance the *Campanula Erinus*, the Rovo, the fruit of which is called the spotted mulberry, the plantain, the agnus castus, with the sweet melissa and the graceful *Torilis nodosa*. The *Similax aspera*, and the *S. Mauritanica* with their climbing stalks cover the rocks and heaps of stones."

"Common in such rural spots are the wild plum, quince, pomegranate, and German medlar trees. There also we may notice the *Celsia cretica*, the *Diphlotaxis viminea*, and a goodly number of grasses. On the coast grows the *Cakile maritima*, and every where in Gozo we find tufts of *Ononis ramosissima*, which is dried and used in the caulking of ships."

"The *Crucianella maritima* exhales its balsam-like perfume at eventide, and every spring gives life to beds of water cress. At San Paolo di Marsa el Forn the botanist will find the striped convolvulus, and upon the lofty Ta Cenc Rocks the sturdy *Euphorbia dendroides*. Upon the hill Ta Harrax luxuriate the *Conixa Saxatilis*, the *Lotus cytisoides*, the tall *Ferula nodiflora*, and the *Ruta chalepensis*. In the fields which flank this hill grows amongst the standing corn the *Bartsia trixago* var. *versicolor* and under the stone heaps by the sides of these fields *Helmintia echoides* in great profusion. In certain waste places we meet with a variety of the *Cardus pycnocephalus* of *Linnaeus*, which differs from the typical species in having perfectly white flowers.

Since the Chair of Natural History and Medical Jurisprudence has been occupied by Dr. Gavino Gulia, the Botanical Garden has been re organised, and now contains some 2000 foreign plants. The most critical and rare plants belonging to the Maltese flora are also cultivated, so that the tourist lover of botany will have no need to visit the ravines during the great heats of summer.

CHAPTER III.

NATURAL HISTORY.

Maltese Naturalists.—Wild animals.—Marine Mammalia.— Domestic animals.—Maltese Dogs.—The Economico-Agrarian Society.

AMONGST the careful observers of birds, beasts, reptiles, and zoophytes may be mentioned Sig. Antonio Schembri who in 1843 published a valuable list of birds observed in Malta and Gozo, and in January 1864 there appeared in the columns of the Ibis a most careful and able enumeration of 253 different species by C. Wright Esqre., since which time this eminent naturalist has added 14 others, making a present total of 267 species. W. C. P. Medlicott, Esqre., W. Grant Esqre., Dr. A. L. Adams, the Revd. H. Seddall, Dr. Gavino Gulia, Sig. G. Mamo, Sig. Gaetano Trapani, Dr. A. A. Caruana, Mr. Davidson, and others have also studied the Maltese fauna, and the student of Natural History is referred to their works which can be obtained from the local libraries and booksellers.

The mammalia of the Maltese islands are represented by a few well known European forms. The wild animals are the weasel, hedgehog, rabbit, Norway rat, several species of mouse, and the horseshoe and long eared bats.

The weasel is the solitary representative of the carnivora and is a determined foe to the rabbit. It is seldom seen, living as it does in dikes and stony places. The hedgehog prefers the cultivated districts as does also the rabbit. This latter animal was formerly strictly preserved at certain places such as Corradino Hill, the neighbourhood of Fome-rieh, the island of Comino, etc., by the Knights, no Maltese sportsman being allowed to shoot or otherwise destroy it. It is said that in less than seven years Sir H. C. Ponsomby when Governor of Malta had 11,000 rabbits killed off near Marfa. Rabbits in Malta are less warmly clad in fur than their English family connexions. Marl heaps along the shore are convenient for burrowing purposes, but men and weasels are terrible foes to them.

The little horse-shoe bat *(Phinolophus hipposideros)* is often seen during the summer, and is sometimes tempted forth by a mild day in winter. The long eared bat *(Plecotus communis)* which has relatives both in Southern Europe and Northern Africa, finds a home in the caverns and catacombs of Città Vecchia. The Norway rat and common mouse are certainly by no means extinct.

The marine Mammalia are the Monk Seal (Phoco Monachus), the porpoise, and one or two species of the whale tribe. Specimens of this cetacean are occasionally found stranded, as was the case not long ago near the General's Rock in Gozo. The dolphin *Delphis tursio* is abundant but being naturally timid does not often approach the shore. The fishermen meet with it at a distance of some six miles from the coast.

Mules and asses which are employed to tread out corn, and are yoked with cattle in the fields, are of large size, and the Knights of St. John set great store by a superior breed of these animals, called janets, which were formerly often exported to America and elsewhere. A few specimens worth about £20 may still be seen in Gozo. The live stock, including some 6000 cattle number about 25,000. There are two well marked sorts of cattle. One is a large, fawn-coloured, bony animal, which was evidently formerly very powerful, though at present of a very degenerate type. The cows, which are frequently used in the cultivation of the soil, give but little milk and often produce two calves at a time.

The Barbary ox is a smaller animal, which is generally imported from Africa in a lean condition, but after being stall-fed for a short time on green barley and clover or at other seasons on bruised pulse, or barley mixed with bran, and plenty of the fat producing cotton seed, he quickly gains flesh, and is speedily converted into beef. At certain seasons these animals are fed entirely on the leaves of the prickly pear which gives a very peculiar flavour to their flesh. The horses of Malta which are barbs, are being greatly improved by the introduction of English and Australian blood. They have the character of giving the colt-breaker some trouble. Sheep and goats are exceedingly prolific, ewes sometimes bringing forth as many as four lambs at a birth. From the scarcity of pasture, mutton is not of the best quality, and both sheep and goats have greatly degenerated. They are the chief milk-producers, a good goat giving as much as two quarts at a time. In Valletta the goat-herd leads his bleating animals, whose udders are of unusual size, from door to door for the supply of customers. Cheese is made from the milk of the sheep, more especially in Gozo.

The Greeks and Romans set great store by Maltese dogs, and Aristotle describes them as being small but beautifully proportioned. Timon says that the Sybarites used to take little Maltese dogs with them when they went to the bath. These animals had long silky hair, and are described by Buffon under the name of Bichons, as being a cross between the small Spanish terrier and the little barbet. Malitheus after Aldrovandi who wrote a very good description of this species gives it the name of

Canis familiaris Meliticus. It is now almost if not quite extinct. It was remarkable for the very slight affection which it evinced for its master. Plenty of loudly barking dogs are however still to be met with, but several cases of hydrophobia having been reported the police have of late cut short the life of many a cur.

Sir W. Reid whilst Governor of Malta re-organized the Economico-Agrarian Society for the improvement of the breeds of domestic animals, and for the general advancement of agriculture. This Society has done good service by holding annual Agricultural Shows at the Boschetto on the popular festival of S. S. Peter and Paul (June 29th), at which prizes are given for the best animals and farm produce. The same useful Society also holds an annual Flower Show at the Upper Barracca in Valletta, at the same time maintaining there a pleasant garden which is a deservedly popular resort, and much frequented both by residents and visitors.

CHAPTER IV.

ꞵIRDS

Ornithologists. — Periodical Migrations. — Influence of Winds.—Malta in Spring. — Indigenous Birds. — Feathered Visitors.—Winter Birds.

THE following outline has been gleaned from the" Notes of a Naturalist" by Dr. A. L. Adams, " Malta Past and Present " by the Revd. H. Seddall, and the exhaustive catalogue of Mr. C. A. Wright, to which with other kindred works such as the " Repertorio di Storia Naturale" of Dr. Gulia, Mr. W. Grant's " Birds found in Malta " and the lists given by the Marquis Barbaro-Crispo and Sig. Schembri reference should be made.

Mr. Wright's original list contained:—"*Raptores*, 28 ; *Insessores Dentirostres*, 57; *Insessores Conirostres*, 33; *Insessores Scansores*, 4; *Insessores Fissirostres*, 13; *Rasores*, 9; *Grallatores*, 64; *Natatores*, 47. He has since added 14 other species making a total of 267. He says:—"Only 10 or 12 species are resident, that is remain with us all the year round, Malta

being merely a resting place for birds in their periodical migrations across the Mediterranean. The
arrivals of birds chiefly take place at the period of the
vernal and autumnal equinoxes." The ornithologist
must be on the alert from the middle of March till the
end of April, and also during the autumn migration
which is known as " the great passage." " Occasional
visitors appear during the winter months, and a
few in summer. Birds generally arrive and leave
at night, and do not usually remain more than one
day, thus giving little opportunity of studying their
habits. Some species however remain a few months
on the island, and several of them breed here
en-route for Europe, whilst flocks chiefly of Grallatores
and Natatores may be seen passing high in the air
without alighting."

"The influence exerted by the wind over these
migrations is no doubt very great. In spring, the
quails and most of the short-winged and smaller
birds, and such as are of weak flight, though not unfrequently arriving in calm weather, generally appear
during the prevalence of winds from the N. N.
W. to S. S. W., and in autumn with those from
the S. S. E. to N. N. E., being probably stopped
in their migratory course, and driven to seek rest
on our shores." Dr. Adams is of opinion that a
strong sirocco or north wind and dense sea-haze
cause the quail to come upon the island unawares.
"Sometimes a fresh breeze springing up from any

point will bring with it numbers of the smaller brids; and if it increase in strength the larger birds and those of stronger flight will also make their appearance. But there appears to be no rule for birds strong on the wing, which arrive under all circumstances of weather and with winds from all points of the compass."

Mr. Wright continues: " It is more especially in spring that in the rocky "wieds" we find the bright coloured Bee eaters, Orioles, and Rollers sheltering themselves from boisterous winds, while the dense foliage of the Carob trees and Orange groves serve them and many others for shade and roosting places. At this season the Harriers scour the rocks and corn fields; the Quails crouch amongst the tangled stalks of the crimson Sulla; the Larks, *(Alauda brachydactyla)*, hover over the rocky wastes, covered with the aromatic *Thymus Capitatus*; and the numerous thickets of prickly pear (*Cactus opuntia*), fig, and pomegranate trees provide resting places for Warblers. The air is perfumed by thousands of wild flowers ; here and there rises a tall palm ; and the Arab houses, language, and origin of the inhabitants indicate, despite Acts of Parliament and a European fauna, Malta's alliance with Africa and the East."

The Revd. H. Seddall says that the indigenous species comprise only the Jackdaw, which breeds iu cliffs and the fortifications of Valletta; the Blue

Solitary Thrush, a lover of rocks and solitude; the Spectacled Warbler, generally to be found in the Military Cemetery at Floriana; perhaps the Robin, the Herring Gull, and the Kestrel, which breeds in cliffs and fortifications. To these may be added the Barn Owl, which breeds in ruined walls near Valletta and the Three Cities, the Rock Pigeon, which rears its young on the southern shores and at Filfla, with the Cinereous and Manx Shearwaters, and the Storm Petrel, which also select the same localities for their domestic establishments. Fort Manoel Island, the Marsa, the Salini, Marfa, etc., are spots loved by the ornithologist.

The Egyptian Vulture is a rare visitant, but the Imperial, Spotted, Short-toed, and Golden Eagles, together with the Osprey are sometimes shot. The Rough-legged and Honey Buzzards feed on lizards and small birds. The Common and Black Kites are very rare. The Maltese call the Marsh Harrier, Bu-Ghadam or "the father of bones," naming the Hen Harrier Bu-ghadam abiad or "the white father of bones." Montagu's and the Pale-chested Harriers are also seen in March and September. Numerous hawks pause in Malta during the spring and autumn, such as the Sparrow Hawk, Little Red-billed Hawk, *Falco barbatus*, the Peregrine and Eleonora Falcons, the Lanner, Goshawk, Hobby, Orange-legged Hobby, Merlin, Kestrel, and Lesser Kestrel.

Of the family of the Owls we have the Barn Owl, which breeds in the battlements of Valletta and the Three Cities, with the Sparrow, Scops, Short, and Long-eared varieties. The Wryneck locally known as "the King of the Quails," or "the Father of Crouchers" is one of the earliest visitors in spring and autumn, at which seasons the Cuckoo is also common. The Great Spotted Cuckoo is very rare, as are also the Crossbill, and Bullfinch.

The Vinous and Scarlet Grosbeaks are winter visitants, and we must not forget the Serin, Green, and Hawfinches, or the Spanish, Tree, and Rock Sparrows. The Finches are the Chaffinch, Bramble Finch, and Siskin, which is often crossed with the Canary by the native bird-fanciers. A few Linnets breed in Malta, but of the Buntings the Cirl, Meadow, Reed, Snow, and Black-headed varieties are rare, whilst the Common and Ortolan species, the latter of which loves to bathe in rain water upon the rocks, are not uncommon. The beautiful Golden-crested Regulus is somewhat rare, as is also the Fire-crested species. The Rook pays us a visit, the Jackdaw is a resident, the Magpie and Sardinian Starling have been shot, and the Common Starling is numerous during the autumn and winter. The Rose-coloured Pastor arrives at irregular intervals.

The Swallow and its relatives must not be passed over. They are the Common, Rufous, and Rock Swallows, the latter species being perhaps resident in Gozo. Also the House and Sand Martins, with the Common and White-bellied Swifts. The Night-jar, *Caprimulgus Europœus* is shot and snared in large numbers for the table, but the Rufous-necked Goat Sucker, *Caprimulgus Ruficollis* is rare. The Spotted, Pied, and White-necked Fly-catchers arrive and depart in spring and autumn, and we see the Grey, Great Grey, and Lesser Grey Shrikes; the Woodchat Shrike known as Bu-Ghiddiem or the "Father of Biters" being much more common. The Red-backed Shrike is not unknown.

The Skylark and its congeners, the Crested, Wood, Short-toed, Cream-coloured, and Calandra Larks are more or less numerous, whilst of the Pipits we have the Richard's, Tawny, Meadow, Red-throated, Tree, Rock, and Water varieties, together with White, Grey, and Yellow Wagtails.

The Golden Oriole visits the islands regularly in spring. This most beautiful bird is exceedingly fond of the fruit of the Nespoli or Japanese medlar. The Blackbird, Great Titmouse, Ring Ouzel, Song, and Missel Thrushes swell our list of birds at the spring and autumn migrations. A few Fieldfares are caught every year in January, and the Redwing appears at irregular intervals. The Rock Thrush pays us a *flying* visit twice yearly, and next to the Nightingale the indigenous Blue Solitary Thrush is prized for its song. It is remarkable for its attachment to the locality in which it has

been brought up. High prices are often paid for
good songsters, and the Maltese often suspend a
piece of red cloth and a cowry shell in the cage,
which they consider a certain specific against the
"evil eye"!

The Common, Russet, and Eared Wheatears
visit us, as do also the Winchat, and Stonechat. The
Nightingale, together with numerous Warblers and
other small birds is taken in nets which are thrown
over a low spreading carob tree selected for the
purpose, the birds being driven from other trees
into it.

The Redstart together with the Black variety,
and the Robin deserve mention. The Hedge Sparrow
is rather rare, as are also the Blue-throated Warbler,
and Blackcap. The Garden Warbler is sometimes
brought to market to the number of a hundred
dozen at a time. It is the far-famed *beccafico* of
the Italians. We can only enumerate the White
Throat and Lesser White-throat, with the Orphean,
and Subalpine Warblers. The Spectacled Warbler
is a resident, and haunts the Military Cemetery at Floriana. The Maltese call the Black-headed Sardinian
Warbler *Ghasfur tal maltemp* or "the bird of bad
weather." The Dartford, Willow, Wood, Bonelli's
and Vieillot's Willow Warblers are all found upon
these shores.

The Chiffchaff bears the name of Bu-fula, or
" the Father of a Bean." The Sedge Warblers are

three in number, viz., the Common, Rufous, and Great varieties. The Reed, Savi's, Moustached, and River Warblers are all catalogued by Mr. Wright. The Hoopoe which is said to breed in great numbers in Tripoli, the Roller, a bird which occasionally makes a nest in some ruin, and which Englishmen sometimes call "the Blue Jay," and the two Bee-eaters, which are shot by the score at one discharge, are amongst the most beautiful of our feathered visitants.

It is said that the Kingfisher sometimes breeds here, but the Wood and Stock Pigeons do not. The Rock Pigeon rears its young in considerable numbers on the southern shores of the islands, and also at Filfla. The Turtle Dove, strong on the wing, is caught in large numbers in platform nets by the aid of the hooded decoy birds. For a full account of this sport see Mr. Wright's list before referred to. Dove catching is a thoroughly clerical amusement in Malta. The natives are good marksmen, and are very skilful in luring birds by imitations of their notes. The Pintail Sandgrouse comes now and then, and the Quail is the principal game of the sportsmen of Malta. Fifty or sixty brace may be shot in a day, but ten or fifteen brace are ordinarily a very good bag. Some quail are also caught by imitating the call-note of the female, and so drawing the males, which are the first to arrive, into nets spread on the standing corn.

The Bustard, Little, and Ruffed Bustards are rare, as is also the Cream-coloured Courser which the natives call the English Plover, but the Thick Knee may be almost considered resident. The Golden Plover, Dotterel, Ringed, Little Ringed, and Grey Plovers are common, and visit the islands regularly, but the Kentish, White-tailed, Asiatic, Golden, and Spur-winged Plovers are more or less rare. The Oyster Catcher is only an accidental visitor, but the Collared Pratincole and the Lapwing arrive and depart year by year.

Common and Numidian Cranes are sometimes seen, and Common, Purple, White, Squacco, and Buff-backed Herons, together with Egrets, pass and repass, as do also the Bittern, Little Bittern, and Night Heron. White and Black Storks, and the Spoonbill are alike rare, but the Glossy Ibis, Curlew, Whimbrel and Slender-billed Curlew, are regularly seen and shot during the spring and autumn. The Black-tailed and Bar-tailed Godwits are not common, but the Greenshank, Redshank, and Spotted Redshank are more so.

Sandpipers are often observed. We may mention the Marsh, Wood, Green, Common, Bartram's, and Curlew varieties, together with the Sanderling, which last occasionally finds its way to our shores. Ruffs, Great, Common, and Jack Snipe, as well as Woodcocks are fairly abundant at the periods of migration. The Knot is rare, but the Dunlin and

Stint are common, though Temminck's Stint is much
less so. The Turnstone and Avocet are somewhat
rare, but the Stilt is more frequently met with.
The crimson-winged Flamingo which is so abundant
upon the inland waters of Barbary is merely a accidental visitor. The Water Rail which the Maltese
call the Winter Rail is not very common, but the
Corn, Spotted, Baillon's and Little Crakes are in
some years fairly abundant. The Water Hen, Coot,
and Crested Coot, are not uncommon, nor are the
Skua and Pomarine Skua quite unknown.

Of the Gulls the Lesser Black-headed, Herring,
Audouin's, Common, Kittiwake, Slender-billed, Adri-
atic or Mediterranean, Black or Brown-headed, and
Little varieties hover above and around us. The
Terns are represented by the Sandwich, Common,
Lesser, Black, Whiskered, and Gull-billed species.
The Cinereous and Manx Shearwaters which are
amongst our indigenous birds rear their young among
the cliffs of the southern shores of Malta and Gozo,
at Filfla, and at Comino, allowing themselves often-
times to be taken whilst sitting on the nest. To
the Storm Petrel we have already alluded. The
Cormorant and Pelican are by no means regular in
their visits, and the Bean Goose seldom makes any
stay with us, usually flying high over head, as do
also the Hooper and Mute Swans. A flock of the
last named noble birds was however seen in the
Quarantine Harbour on December 23rd, 1865.

The Common and Ruddy Shieldrakes head the list of the genus *Anas*, of which the Shoveller is one of the most common, the Mallard being also a winter visitor. The Pintail Duck and Gadwall come only occasionally, but the Widgeon, Teal, and Summer Teal are seen in larger numbers. The Pochard, and the Tufted, Red-crested, Whistling, and White-headed Ducks are scarce, but the Nyroca Duck is perhaps the commonest Duck that visits the islands. The Red-breasted Merganser, Smew, Red-throated Diver, the Crested, Horned, Eared, and Little Grebes, with the Guillemot and Puffin close our list.

Dr. Adams says:—" To the ornithologist there is not much variety in the fields in midwinter. Among the crops of cacti, (*C. Opuntia*), a solitary song thrush or blackbird is occasionally seen; from the dike-top the ringing note of the bunting, (*E. milaria*), the chirpings of the reed sparrow from the house-top, robin, and the chiff-chaff utter their well-known call-notes. A few song larks, and a solitary pied or grey wagtail are occasionally observed; but of all the midwinter tenants of the fields the tit-lark is the most plentiful. A stonechat, or the white-fronted redstart hops along some stony lane, whilst small flocks of chaffinches are seen among the tree tops. About this season of the year, when the northern blasts blow strong, and the gregale lasts for three days at a time, there may appear such accidental visitors as the fire and golden-

crested wrens, pelican, crossbill, fieldfare, missel-
thrush, rook, etc; but many of the early winter
birds push southward by the middle of Jannuary,
as soon as the fields have been ploughed, and the
crops are getting up. If there is one pleasant re-
miniscence more acceptable to my memory than another, it is those happy winter days, when I used
to crawl along the beetling cliffs of Emtahleb and
its neighbourhood fossil-hunting, with the blue
thrush, serin finch, linnet, and spectacled warbler,
singing sweetly among the olive trees below me. "

CHAPTER V.

REPTILES.

DR. GULIA has published a complete list of the reptiles existing in Malta in Il Barth, a medical and scientific journal of which he was the editor. The common Land Tortoise or Fecruna ta l' ard is not used as food, although its flesh is palatable, The common Turtle or Fecruna tal Bahar is often taken by fishermen.

Lizards are plentiful enough especially the Wall Lizard or Gremxula, which changes its colour with ease. Strange to say, all the lizards on the islet rock of Filfla are of a beautiful bronze black, a colour not to be found on the mainland. The *Gongylus ocellatus* or Ocellated Skink, which the natives call Xahmet l' ard or " the fat of the earth" is a smooth, slippery, fat reptile, with a skin like a snake's and very short legs. It grows to the length of 8 or 10 inches, and lives under large stones. The *Ascalobotes Mauritanicus* or Italian tarantola is called in Malta Uizgha Scuda, and the name of Uizgha is given to a small ugly house-infesting lizard. The

bones of a large fossil lizard were found by Dr.
Adams at Benghisa. Only two species of snakes
are indigenous, both of which are quite harmless.
They are the *Coluber viridiflavus* which the Maltese
call Serp or Ghul, and the Spotted Snake or Lifgha.
These snakes are plentiful but timid, sometimes
attaining a length of 23 inches. According to native
tradition St. Paul banished all venomous snakes
from Malta, as St. Patrick did from Ireland, and
the saliva of persons born on the festival of the
Conversion of St. Paul is said to be efficacious in
the cure of snake bites, as are also St. Paul's earth,
and the *Terra Sigillata Melituæ.*

The Painted Frog which croaks in numbers in
the pools, in the aqueduct at the Marsa, and sometimes in brackish water was formerly fried and eaten
on fast days, and was also given in the form of
soup to sick children.

In addition to the list of reptiles by Dr. Gulia
already referred to, see also Dr. A. L. Adams'
"Notes of a Naturalist," and "Malta Past and
Present."

CHAPTER VI.

MALTESE ICHTHYOLOGY.

Books of Reference.—Methods of Fishing.—Finny Residents and Visitors.—Fishery Regulations.

The following remarks are for the most part translated and abridged from the admirable "Tentamen Ichthyologiæ Melitensis" of Dr. Gulia, who has most kindly given me every possible assistance. Carsten Niebuhr a Danish naturalist was the first to publish a list of Maltese fishes in 1775, compiled by Giorgio Locano, which described some 116 species, and in 1838 Sig. Gaetano Trapani compiled a catalogue in which he enumerated 157 different species. Dr. Gulia in 1861 made mention of 186 species, belonging to 108 genera, and 47 natural families. The markets should be visited in the early morning, and at the fishing port of Migiarro in Gozo many curious specimens may be met with. Fish are fairly abundant and cheap. Their colours are more gorgeous, but their flavour is inferior to those caught in more northern latitudes. The hand line *lenza*, and the wicker pots *nasse*, the large seine

xarpa, and the small seine *tartarun,* together with
the trammel *parit,* the casting net *teriha,* the long
handled fish spear *foxna;* and the rod are all employed by the fishermen. The latter is most successful
after a gregale. Fishing with a white feather below
which is a hook whilst sailing briskly is often amp-
ly rewarded. The Lampuca a large species of
Mackerel is caught by this means, and also by long
lines with almost countless hooks, cuttle fish being
used as bait.

The Sea Lamprey or Kalfat famed for activity
is seldom caught, being deficient in flavour. Malta
can boast of two varieties of the Muræna (Murina),
of exquisite flavour, but of which the bite is dan-
gerous, viz., the very abundant and much prized
Muræna hælena or Yellow-spotted Eel, and the *Mu-
raena unicolor* which is extremely rare. The Common Eel which the Maltese call Sallura is plentiful. Four species of Conger Eel or Gringu inhabit
these waters. They are the Rock, Sand, White, and
Black Congers. The Sea Viper called here the Sea
Snake has good but indigestible flesh. Of these
gluttons of the sea the two former are edible, the
second being generally preferred.

Passing on to the family of the *Clupeideae,* we
note their commercial importance. The Anchovy is to
the Mediterranean what the Herring is to the northern waters. This fish known in Malta as the In-
ciova is gregarious in its habits. With the excep-

tion of the Bitter Anchovy the various species are deservedly esteemed for the table. The Sardine, shoals of which are caught in the fishermen's seines must not be forgotten. The *Salmo Fario* or Salamùn is but rarely caught, and two species of fish which Dr. Gulia classifies as *Micromugil timidus* and *Micromugil macrogaster* are found in the aqueduct at the Marsa, and round the shores of Malta.

Five species of the *Scomberesocidae* are met with, amongst which are the Flying Fish or Rondinella, the Sea Pike or Litza, and the Needle Fish or Imsella. The Maroon or Ciaul is a favourite food of the feline race.

The rock-haunting Wrasse or Tirda which belongs to the family of the *Labridae* often repays the patience of the angler. Dr. Gulia has done much to classify the different species of Wrasse, and to his most interesting treatise the lover of ichthyology is referred. We must not fail to mention the *Scarus Creticus* or Martzpan, renowned of old in story, which Epicarmus in the fifth century before the Christian era, said was a dish for the Gods. From its beautiful colours it has been styled the Parrot of the Sea. It was formerly somewhat rare on these shores, but has of late been plentiful.

Of the *Acantini* we have ten or eleven species, most of which are edible. The Galera or Ballottra tar-ramel, is the *Ophidium barbatum*, and is the only species of this genus found in our waters. It is neither very abundant nor highly esteemed for the table.

The *Gadideae* or Cod family are somewhat numerous, and are represented by some four or five varieties. These are the Pecorella or Ballottra, the Merluzzo or Marloz, and the two varieties of the Lipp or Sea Tench. The famous Remora or Pesci Tmun, with its curious sucker must be included in our list.

Of the *Pleuronectideae* or Flat Fish Dr. Gulia enumerates four species as belonging to the fish of Malta. The *Rhombus Laevis* or Barbun is the most plentiful. The *Rhombus Maximus* or Turbot is very rare. Two kinds of Sole are also caught, the common variety being styled by the natives Linguata. The Order of the *Acanthopteri* furnishes many a dainty for the table.

The *Trachini* or Weever Fish, which it is said derive their name from their tenacity of life, are represented by three varieties. They are edible, but are justly dreaded by the fishermen on account of the wounds inflicted by their dorsal fins. Hence the great Weever is often called by English fisher folk the Sting-Bull. The *Percideae* or Perch family are numerous. The Basse or Spnotta is rare, but is a very handsome fish. *Apogon rex mullorum* or Beardless Mullet which is much esteemed by eaters of fish, really belongs to this family. The Rock Cod or Cerna, and the Thorny Perch or Hanzir grow to a large size, and inhabit deep water, whilst other varieties are caught not far from shore. The Sea Perch is highly recommended to convalescents as being easy of digestion.

The Red and Yellow-striped Mullets or *Triglie*, the former of which loves the rock, whilst the latter prefers the mud, are well known fish. Six species of the mullet are mentioned by Dr. Gulia. Of the *Triglidiae* eleven species swim in our seas, including the beautiful Flying Gurnard, called by the Maltese Falcun or Bies.

The Sow-fish and Sea Scorpion are nutritious, but another variety of the same tribe which inhabits deep waters is but lightly esteemed.

The family of the *Sparideue*, can boast of 17 species, amongst which we may especially note the Vopa or Boops. The fishermen love the fish of this family well.

The migratory *Scombriedae* or Mackarel which are plentiful at all seasons muster in great force, and are of immense value to the dwellers on the shores of the Mediterranean. The Pilot Fish is foolish, and tenacious of life, but no delicacy. The four varieties of the Tunny are with one exception deservedly and highly prized, and the Sword Fish, albeit of no great size is not to be despised. The Lampuca is an autumn visitant most welcome to the fishermen, but the oily Turkish species is rare, as it prefers Sicilian waters. The Bonito is a great delicacy, but the Plain or Striped variety is rarely eaten, nor is the Horse Mackarel highly thought of. The John Dory or Pesci San Pietru deserves mention.

The *Cepolideae* are represented by the Red and White varieties of the Snake Fish, and by *Trachypterus Spinolœ* which last was caught at Marsasirocco in 1871.

The Blenny tribe is not of any great value as food, and two varieties of the Sea Devil or Petricia are caught, together with two several kinds of Hippocampus or Sea-Horse.

The *Plagiostomi* are numerous. One of the varieties of the *Scyllideœ* is the Catfish, which is very plentiful, but its flesh has a disagreeable odour, which is only partially removed by being steeped in water. The poorer classes eat it, and its native name is Kattus. The Lesser Spotted and the Black Mouthed Dog Fish are caught.

The Pesce Cane or Penny Dog which whilst young is called the Miller Dog is common, the female producing from 60 to 80 young every year. The Blue Shark which the Maltese style the Sea Dog is still more ferocious.

The White Shark is the most terrible of his tribe, but is fortunately a somewhat rare visitant.

The *Lamna cornubica* or Smeriglio which sometimes attains the length of 24 feet is edible, but woe to any hapless mariner on whom his cruel jaws may close!

The Sea Fox or Fox Shark is rare and of singularly cunning habits. The Grey, Smooth, and Picked Sharks may be mentioned, together with the Shark Ray, Balance Fish, and Saw Fish.

The Rays nine in number, such as the Cramp, Thornback, Spotted, Sharp-nosed, Sting, and Eagle varieties are voracious, and lovers of darkness. They have been compared to the birds which seek their prey by night. Edible themselves, they are terrible fish slayers.

Fishery regulations and a close season have lately been established, to the no small benefit of the fisheries, which were fast becoming impoverished.

CHAPTER VII.

Entomology.

Beetles.—Cockroaches, Locusts, and Grasshoppers.—
Dragonflies etc.—Bees, Wasps, and Ants.—Butterflies and
Moths.—Parasites.—Flies and Gnats.—Scorpions, Spiders,
etc.

IT was reserved for Dr. Gulia to publish the
first, and as yet, the only standard work on insects
found in Malta and Gozo. Mr. Leach in 1832
collected numerous specimens, which he sent to
the London Zoological Society, but his lamented
death from cholera at Genoa prevented the publi-
cation of his intended work. In 1857 under the
auspices of Sir W. Reid, who was then Governor
of Malta, Dr. Gulia delivered a course of entomo-
logical lectures to a class of students at the
Palace of St' Antonio, which he afterwards published.
These he has most kindly allowed me to use, and
from this source the following pages are drawn.
Three varieties of the Tiger Beetle are met
with, of which *Cicindela littoralis* and *C. hybrida*
love the 'sand, whilst *C. germanica*, which is only
half the size of the other two, prefers herbage.

Passing on to the *Carabidœ* we have the voracious
Calosoma sycophanta, *C. indagator*, *Carabus granulatus*,
Carabus laevigatus, *Procrustes coriaceus*, and numerous
varieties not as yet classified.

In stagnant pools dwell *Gyrinus fontanalis* and
Dytiscus circumflexus. Of the Burying-Beetles, known
in France as *Boucliers* we have two varieties *Silpha
sinuata* and *S. obscura.*

Several species of the Scarabaeus form part of our
list, of which *Bubas bison* is not very common. The
name of Bukuar is locally given to three species
collectively. The Horned Beetle is the largest
insect of the order which we possess, the Stag
Beetle is rare, but the Tumbler Dung Beetle abounds
in the fields, and the Cockchafer must not be
excluded.

The Barbary-bug or Busuf is terribly destructive
when fruit-trees are in flower. The Rose-chaffer is
here called Ghaur or the Digger, and we have also
the swift Rove Beetle. *Buprestis tenebricosa*, *B. dis-
coidea*, with *B. viridis* and several other species are
well known, as is also the Glow-worm or Musbih
el-leil.

The Darkling Beetle or Hanfusa together with
a smaller variety is found in moist spots, and the
Field Beetle or Hanfusa Tar-raba is everywhere
plentiful. The Meal Beetle is a pest to millers and
storekeepers, and at least two species of Blister
Beetle, one of which is found on the blossoms of

the Chrysanthemum, have been observed. The Soft
Beetle or Dliela endangers the lives of animals that
inadvertently swallow it. Pea, Rice, and Grain Weevils
are unwelcome guests, the *Lixus parapleticus* is
hurtful to horses when swallowed, and *Brachycerus
undatus* and *barbaṙus*, of which negro women make
necklaces and amulets are common.

The Golden Beetle specially loves the cat-mint,
and we have also the variety called in France the
Gilded Harlequin, not to mention others. The *Crio-
ceris asparagi* has been seen in Gozo. Lady-birds
called Dud ta l'iscola are plentiful.

The *Orthopteri* those scourges of industry abound
in these islands. We have the Black, Red, and
German Cockroaches, the latter being small and
rare. Three species of Mantis or Walking Leaves
bear the name of Debba ta l'Infern. Two of them
are *Mantis oratoria* and *Mendica*. Red, Blue, and
Green Grasshoppers swarm, and the Migratory Locust sometimes threatens us with its destructive
visits. In 1850 a swarm passed over the eastern
portion of the island, doing damage at Casal Zab-
bar and at Wied el Ghain, and covering the sea
with their bodies. The Mole Cricket is also very
hurtful to the crops, the Field-Cricket proverbial for
stupidity utters its shrill note in summer, and we
have two varieties of the Long-headed Grasshopper
and the Ear-wig.

The *Neuropteri* are here named Mazzarelli. May-flies, Dragon-flies, two varieties of the Ant-lion, *A-grion puella*, *Calepterix virgo*, with one species of *Œshna*, *Lestes*, and *Chrysopa*, sum up this class of insects.

The Wild-fig and Oak-leaf Flies which are better known as Gall-Flies, and the Ruby-tailed Fly all dwell in Malta. The female *Urocerus Gigas* stings sharply. The Blue Bee or Nahal Baghli, the Spot-ted Bee, the Mason Bee or Nahal Bennei, the Hive Bee or Nahal, 'the *Nomada bi-fasciata* and *N. uni-fasciata* are all Maltese insects, as are also several kinds of wasps, such as the Sand Wasp or Baghal, *Vespa Ichoris*, *V. Grecorum*, and *Polistes gallica*. *Sphex spirifex* is fond of grapes, and *Scolia flavi-frons* is very common.

To the Ant the Maltese give the name of Ne-mel. Two kinds of Red Ants, the Turf Ant, and *Formica herculanea* dwell here. The winged Ants are called Sultan el nemel or King of Ants.

Caterpillars have the local name of Duda tal Farfett, the chrysalis being styled Fosdka, and the butterfly Farfett. The Cabbage and Turnip Butter-flies are common, and in the caterpillar stage do much mischief. We may enumerate *Rhodocera Rham-ni*, *R. Cleopatra*, *Vanessa Atalanta*, the Swallow-Tailed Butterfly, here called the Rue Butterfly, or Farfett tal Feigel from its preference for that plant, the Queen's Page, or *Papilio Podaliurus*, Ochsenhu-

*meria Cardui, Colias Edusa, C. Lesbia, Polyomat-
us Phlœas,* and the beautiful *P. Argus,* which the
Italians call Hundred Eyes.

The Humming-bird Hawk Moth or Bahria is
sometimes called the Pigeon's Tail Sphinx. The
Death's Head Moth is called Farfett el Meut or
Ras el Meut. The Humming - bird Hawk Moth is
looked upon as the harbinger of bad news, whilst
the sight of the two varieties of the Red News-
monger is considered a happy omen. The Convolvulus Moth feeds upon the Convolvulus whilst in
the Caterpillar stage.

A solitary specimen of *Saturnia pavonia major*
has been caught. The Mulberry Silkworm called
Dud or Farfett tal harir has been reared here for
centuries, and Sir W. Reid when Governor introduced
the Palma Christi species called Duda tal Harir
ta Riccinu.

Clothes Moths are destructive, and other species
spoil quantities of grain. *Deiopeia pulchella* is seen
at eventide.

Two kinds of cicala feed upon the sap of
trees and plants, and *Aphidae* or Brighet-tas-sigiar
are numerous on the centaurea, bean, oat, cabbage,
etc. Indeed every plant seems to have its parti-
cular aphis. Scale insects including the cochineal variety are likewise not wanting. The latter however does
not thrive in Malta. Would that we could say the
same of Black, Green, Wild, and Bed Bugs, with

other pests such as Flies and Gnats ! The Common
and Meat Flies, the former of which is called
Dubbien and the latter Dubbiena tal Laham, have
many friends such as *Musca meteorica*, *M. stercoraria*,
M. pumulionis, and others. Gnats, Sand-flies, and
Gad-flies tease and sting, the latter depositing their
eggs beneath the skins of sheep and oxen, whilst
the Conops or Horse Wasp called Xidia is troublesome
in damp weather. The usual parasites infect men
who love not soap and water, and badly tended ·
animals. Although not coming under the head of
insects, we may here make mention of the Scorpion
or Imkass, from one to two inches in length found
under stones in the valleys, but which never seems
to harm any one, together with the Water Scorpion
or Imkass ta l' Elma, and also the Scolopendra or
Xini Esfar, the Gally-worm or Hanex ta L' Endeua,
the Centipede or Xini ta L' Endeua, and the Wood-
louse or Hanzir ta l' Art.

Numerous Spiders are also to be met with,
amongst which we must not forget the Geometric
Spider, or Brimba tas salib, the Dancing Spider,
or Brimba tal Meut, several kinds of Hunting
Spider, known as Brimba ta Sakajha Tual, and the
Tarantula Spider or Trenta.

The Earth Worm is of course plentiful, but
the Revd. H. Seddall says:—

" Of Tube Worms and other Annelids I have
met with but few species, and these of no remarkable

beauty, with the exception of the *Sabella*, which may frequently be seen on a calm day in the Quarantine Harbour, with its double spiral of tentaculæ projecting from its leathery tube of eight or ten inches in length. They live well in an aquarium."

Much more could be added did space permit, but every entomologist should not fail to study for himself Dr. Gulia's most interesting and able treatise.

CHAPTER VIII.

CRUSTACEA.

THE Crustacea of Malta have been admirably classified by Dr. Gulia and for his list of them see page 314 of the first volume of "Il Barth." The Revd. H. Seddall also enumerates numerous species in an appendix to his book " Malta Past and Present," but it is from Dr. Gulia's list that, with his kind permission and assistance the following facts have been gathered.

The Long-legged Spider Crab about an inch in length is plentiful, as are also the Four-horned and Spinous Spider Crabs. The common Shore Crab loves the mud, and the Swimming Crabs are represented amongst others by the Cleanser, Velvet, and Wrinkled varieties, all of which are useful as food. Land crabs abound in brooks, and in fresh water at the Marsa, Gneina, etc. They are often converted into soup on fast days by poor people, who in consequence suffer from diarrhœa. The Common Pea Crab is very rare, but the Pinna variety is fairly abundant, as are also the Angular Crab, the

Death's-head Crab excellent in soup, the Turk Crab, or Granc tat toroc, and the Sea Cock or Serduk el bahar, a gigantic mollusc of a dark claret colour with an internal shell.

The Hermit crabs number eight varieties, and are plentiful. The *Pagurus Prideauxii* which loves the anemone so well, is not as common as the other varieties.

The Hairy Porcelain crab is abundant, but not so the minute variety. The White and Rough Crawfish, and the Spiny Lobster dear to epicures besides other varieties are common in the markets. The Common and Banded Shrimp, and the Common Prawn must not be omitted.

The *Phyllosomidae* have but one representative, and *Squilla mantis* known as Cicala Baida also stands alone. The Fishing or Shore Worm is used as bait for fish of the Sparus tribe. Its English name is the Sea Slater. Three species of Wood-lice are met with. Several varieties of Water Fleas, three or four kinds of Cowries, one of which *Cypris Pubena* has a horny bivalve shell like a mollusc, and the active *Cyclops vulgaris* may be mentioned.

To quote the Revd. H. Seddall once more. " Many of the crustaceans may be taken with a common dip or landing net from the rocks and quays by drawing it through the seaweed: others by dredging the beds of Zostera in the bays or

harbours. Marsascala and Marsasirocco are good localities. Others can only be taken in the large fish
and lobster baskets called *nasse*, which are laid
down in deep water by the fishermen."

————

CHAPTER IX.

MOLLUSCA.

THERE is in the Public Library a large collection of Malta shells which should be examined by any one who feels an interest in this branch of Natural History. Captain Spratt R. N. and the Revd. H. Seddall were amongst others, zealous collectors, and in an appendix to his well known work "Malta Past and Present," the latter has given much useful information as to the places which will best repay a search. In the first volume of the Medical Scientific Journal "Il Barth" there is (p. 193), the first portion of a classification of the terrestrial and aquatic mollusca of these islands by the Cav. Luigi Benoit and Dr. Gavino Gulia. It is much to be hoped that the remaining portion will ere long see the light.

The late Sig. G. Mamo, in the course of 47 years collected in the Maltese islands and adjacent

seas 438 species of Mollusca. His collection of which
after his death, Dr. A. A. Caruana published a catalogue in 1867, comprised *"Acephala* or bivalves,
145; *Tunicata*, 6; *Brachiopoda*, 9; *Pteropoda*, 8; *Ga-
stropoda*, 259; (of which 42 inhabit land or fresh
water); *Cephalopoda*, 9; *Heteropoda*, 1. This valuable
collection was fortunately retained in Malta, and has
ever since been of great use to all lovers of Nature,
Sir William Reid, who was then Governor of Malta, having purchased it for the small sum of £30,
for the Public Library.

Many Mollusca are here used as food, especially by the poorer classes, such as the *Helix aspersa* known in Malta as *Ghakruxa*. With the exception of the musk-polypus and the paper nautilus, all the cephalapods found hereabouts including
two varieties of the octopus, and several of the cut-
tle-fish are eaten. The gasteropods likewise increase
our supplies of food, and many a univalve is highly
appreciated, as are also numerous bivalves. Marine
molluscs have a more delicate flavour in summer
than in winter, but care must be taken not to eat
those taken from copper sheathing, those which at
certain periods become unwholesome, those which
are of an unusual colour, those which do not shut
spontaneously, and those which have been caught
more than 12 hours in summer and 24 in winter.

The family of the *Muricidae* or Rock Shells
is represented by several varieties, which have a

special affection for the weedy bottom of the Bay of Marsasirocco. *Murex brandaris* is locally known as Sultan el Beccum, or the King Beccum, *Murex trunculus* being styled simply Beccum. Both of these are abundant and edible, and some think that it was from them that the ancients obtained their world famous purple dye. It is however the opinion of Dr. Gulia and other competent authorities that *Purpura Hœmastoma* supplied the dye in question. The edible *Fusus lignarius* is somewhat common, and is called Gharus or Sigromblu tal bàhar. The Maltese call the sea " Bahar."

The *Buccinidœ* or Whelks have numerous representatives, which are found in very large numbers at Marsasirocco, in the harbours of Valletta, and indeed everywhere. Specimens of *Nassœ* are most abundant in the Great and Marsamuscetto Harbours.

Of the beautiful Cone Shells *Conus Mediterraneus* which the Maltese call Sgorra is abundant. *Pleurotoma septangularis* may be met with at Marsascala, Chercheua, and St. Julian's. The rarer *P. multilineata* haunts Marsascala.

Four species of *Mitra* and three of *Marginella* are common upon the sands. *Ciprœa lurida* of the Maltese name is Bahbuha is very common, as is also *C. spurca*, but *C. pyrum* is very rare. Three varieties of Natica are numerous, edible, and known by the name of Ghakrux el bahar or the "sea-

snail." The genus *Cerithium* is represented everywhere, and adds to our food supplies, preferring
shallow waters with a sandy bottom. Its representatives are popularly styled Brancutlu.

The Turret Shells can boast of a *Scalaria*, and
two species of *Vermetus*, one of which bears the
name of Farrett, are tolerably common in the Great
Harbour. *Litorina neritoides* is plentiful all along
the shore, and also upon the stones and sides of
the brackish canal at the Marsa, and several varieties of *Rissoa* will be found at Marsascala, San
Tommaso, and Bir-zebbugia. *Hydrobia ulvœ* is common amongst seaweed, as is also *Neritina viridis*.
The pretty Wreath Shells are also numerous. We may
note *Phasianella speciosa* with its two varieties,
one almost entirely red, and the other milk white.
The Top Shells or Carriers are common both as to
living and fossil specimens. Of the Ear Shells
Haliotis tuberculata is eaten by the Sicilians. Its
Maltese name is Mhara Imperiala. It is found
everywhere attached to stones in deep water. *Jan-
thina bicolor* is rarely taken except at St. Julian's
Bay.

Several varieties of limpets are eagerly collected
for the table, their Maltese name being Mhara,
and the *Crepidula unguiformis* known as Papocc
or "slippers" represents the *Calyptraeidae*, as the
Xifa tal bahar or "sea-thread" does the Tooth Shells.
Chiton cajetanus abounds in the Great Harbour.

Numerous indeed are the *Helicidae*. No fewer than nineteen species of the true snails are met with. The *Helix aspersa* called Ghakruxa ta l' art (earth-snail) or Bebbuxa ragel (male snail) is eaten by the poorer classes. The Moghza or black snail, the ravages of which are justly dreaded by the farmer is a Sicilian dish. The prolific Naghgia or female snail and the egg snail are also wholesome. *Helix gaulitana* is found at Marsa-el-Forn and on the General's Rock in Gozo. Amongst the *Clausilia*, we must mention *C. Delicatae* and *C. Mamotica*, the first of which is only found at St. Paul's Bay, and the second in Gozo. *Physa melitensis* disports itself in fountains and aqueducts, and *Cyclostoma Melitense* is very common under stones in uncultivated grounds.

Bulla *hydatis* commonly styled Baida tal Bahar or Sea Egg abounds in the bays of Marsascala and Cercheua.

Several species of oysters are obtainable, and in 1866 oyster beds were formed in various places, but without much success. The Rev. H. Seddall says "Five species of *Pinna* are found in Malta, some of them common in the harbours within reach of a pole or boat hook. They project from the mud amongst the Zostera roots to which they are attached by their silken cable. Of this silk which is of fine texture, but heavy, I have seen gloves made."

The Mussel, or *Masclu* is found everywhere, and also the *Pholas dactylus* or "sea date." Its native name is Tamla baida or "white date." A Maltese author says that "nature in her bounty to her favourite people of Malta has made even the very stone upon the sea-shore to become pregnant, from whence we draw the delicious sea date."

It is found in soft limestone below water. One of the two varieties has a white, and the other a brown shell. The latter is highly phosphorescent.

Arca barbata called Fardocclu or Spardocclu, found on submerged rocks is the most savoury of its tribe. The edible Cockle or Arzel loves the muddy bottom, five varieties of *Lucinidae*, ten of the genus *Venus* and nine *Tellinidae* may also be noted. Mr. Davidson has described some of the *Brachiopoda*, but many others still await examination and classification. "*Terebratula vitrea* and *T. caput serpentis*, with five species of *Orthis* have been taken in deep water off the islands. Eight species of *Pteropoda*, of the genera *Hyalaea*, *Cleodora*, and *Odontidium*" are included in Dr. Caruana's catalogue of Sig. Mamo's collection, to which, together with other previously mentioned works the reader is referred.

CHAPTER X.

ZOOPHYTES.

THE Zoophytes of the Mediterranean will well
repay the careful student, who will here find a wide
and comparatively unexplored field of labour. No
naturalist has as yet written a full, accurate, and
minute description of the lower forms of Maltese
animal life. Dr. Gulia who has kindly aided me,
has devoted much careful attention to this and
other departments of Natural History, and the publi-
cation of the results of his observations is a thing
greatly to be desired in the interests of science.

Fishermen and others meet with rare and
beautiful zoophytes, but, ignorant of their scientific
value, consign them to the deep, or throw them
upon the rocks to die. About four varieties of
sponge grow in the harbours and in deep water.
It is as yet uncertain whether a fifth variety can
be claimed as belonging to Gozo, or whether it is
but a waif and stray from Sicilian waters. A
gregale brings on shore numerous sponges, seaweed,
and other marine treasures, amongst which we
may note the Portuguese Man of War, *Cestum Vene-
ris* etc.

The Rev. H. Seddall to whose work "Malta Past and Present" reference should be made, says: "Zoophytes of many species are easily found in the rock pools, and growing on the quays of the harbours. Of these *Anthea cereus* is by far the most common, with its long and finely coloured tentaculæ, which are not contractile within the mantle of the animal as in the true anemones (Actinia)."

No fewer than seventeen kinds of anemones may be met with, the scarlet variety of *Actinia mesembryanthemum* being not uncommon round the rocks in sheltered pools. *Sertularia, Flustra, Gorgonia, Celepora, Lepralia, Caryophyllia,* with many other genera are represented. Amongst the *Chelenterata* we must mention *Pelagia noctilica,* which is now and again phosphorescent, especially during a storm and at night. *Rhyzostoma pulmo* must not be forgotten, nor the remarkable *Charybdia marsupialis* so named from having on the abdomen a curious purse, which is by some supposed to be a food receptacle. *Porpita mediterranea* strews the beach at Marsascala after a gregale.

CHAPTER XI.

AGRICULTURE AND PRODUCTIONS.

Maltese Soil.—Cotton, Land Tenure, and Agricultural Implements.—Corn, Sulla, and Potatoes. — Carouba Trees, Fruits and Vegetables.—Figs, Oranges, Grapes, etc.

IT is fortunate for the agriculturist that the Maltese islands are composed of soft rocks which readily disintegrate. Still farming in Malta and Gozo is a battle and victory of labour. The 114 square miles comprised within the insular area are partly barren rock, and in many respects a geographical riddle. Cultivation has asserted its sway over 54,716 acres, the remainder being sterile rock. Owing to the absence of trees and shrubs, the soil of Malta, which is most erroneously said to have been imported from Sicily, contains but little vegetable matter. Plenty of rich alluvial earth which has been washed down from higher levels by the semitropical rains is found in the valleys. In other places the surface rock has been removed, and the fragments built up into walls to hinder the washing away by the rains of the earth, which is found below the surface rock in beds of not more than a foot or eighteen inches in thickness. Many fields have thus been formed and brought under cultivation. Since the English have occupied Malta many more animals have been reared, and land has in consequeuce been more heavily manured. The porous rock below the surface retains a valuable amount of moisture, and heavy dews somewhat supply the want of rain in summer. The cultivation of cotton is thought to have been first introduced by the Phœnicians, and the Carthaginians made Maltese cotton cloths famous in distant lands, on account of their whiteness and substance. Under the Greeks agriculture flourished, the weaving of cotton prospered, and the bee-keepers of Malta were renowned throughout the civilized world. Lucretius Carus and Silius Italicus sing in immortal verse of the fabrics of Malta, Diodorus Siculus speaks of "cloths remarkable for their softness and fineness," whilst Cicero lashes Verres, in no measured terms for his interference with the local traffic in cotton. Under Arab rule agriculture received a check, but on the arrival of the Norman deliverer Count Roger, lauds were re-divided, and the farmer again began to grow wheat and cotton. In 1525, the islands were "unfit to grow corn and other grains, maintaining only a population of 12,000 inhabitants, who lived by exchanging honey, cotton, and the aromatic cummin for the more substantial necessaries of life." In 1687 Gozo produced a large amount of cotton, and wheat and barley sufficient for three months' consumption, importing annually from 7000 to 8000 quarters of grain, and sending many cattle to the sister-island. In 1801 the value of the raw cotton produced in Malta was about half a million sterling. The Civil War in America gave a considerable impetus to the cultivation of cotton, and the crop amounted to more than three hundred tons, but the invention of new machinery and the competition of Egyptian cotton have of late years caused prices to decline. Both the white and dark nankeen varieties are cultivated. The seed is sown at the end of April, the plant flowers in August, and the crop is gathered in September. Cotton thrives well, is sown as a second crop, provides seed for fattening cattle, gives employment to numerous families, and is principally cultivated near Città Vecchia, Zebbug, Siggieui, Zeituu, Lia, Balzan, and Tarxien. Land in Malta belongs to the Church, the Government, and about 2000 private individuals in almost equal proportions. The farms are often let on lease for 4 or 8 years, and sometimes longer. Government leases are usually for a longer period than others, and waste open spaces have been largely sold to tenants for small sums, to be brought under cultivatiou. The implements in use are truly primitive, and may almost be styled Adamitic. Ploughs, which are of the most simple construction, can be carried home on a man's shoulder, and are drawn by cows, asses, oxen, or mules. The

English plough is not suitable for this rocky soil,
and the tillers of the land reject with a smile more
costly and highly finished implements.

The harrow is also of a most original type, and
not unfrequently branches of trees are dragged over
the fields as a substitute. A large hoe is often employed, and in rocky soils the pickaxe is found useful. The land is never permitted to lie fallow, and
the marvellous patient industry of the Maltese farmer overcomes all natural obstacles, and wins from
the soil a return of from 12 to 40 or even sixty-
fold. Wheat, the harvest of which suffices for 3
or 4 months' consumption is sown every alternate
year with barley and clover, in November, and is
reaped in June. Barley ripens a month earlier. Cotton, melons, cummin, sesame, etc., follow. If any
signs of exhaustion of the soil appear, peas, beans,
maize, etc., are substituted for barley. Melons degenerate in Malta. Dr. Gulia introduced the Cantaloupe of Paris with but slight success, but has
succeeded in naturalizing the Cantaloupe of Valpa-
raiso in the Botanic Garden, the second year's seeds
being as good as those originally imported. The
crimson flowering Sulla or clover adds greatly to
the beauty of the landscape. It is the French hon-
ey-suckle or *Hedysarium coronarium*, grows to the
height of from three to five feet, and produces
about 190,000 loads per annum. The Economico-
Agrarian Society, which was re-organized by the late

Sir W. Reid has conferred great benefits on the tillers of the soil. The annual yield of cummin, which is valued at about £2,200, and which is mostly exported, averages from 70 to 80 tons. Malta has two crops of potatoes per annum, which produced some years ago about 800 tons. Increased attention has of late been paid to this useful vegetable. English potatoes degenerate after the second year. The kidney variety has been most successfully grown, and bids fair to supersede all others, as the produce though less in quantity commands double prices. Many sacks of potatoes are annually exported.

Market gardening has somewhat diminished in extent since the appropriation by the public aqueducts of many of the springs, but is still an important branch of industry. The carouba tree is abundant, growing on rocky soil, and its dark foliage is a conspicuous feature in the landscape. It attains a considerable size, and its seed-pods and leaves are used for feeding cattle and horses. Poor people also eat the seed-pods, which when baked, are said to be not unpalatable. Tanks and cisterns are to be found in almost every field and garden, and are indispensable to the gardener or agriculturist. The ornamental plants of Malta are very beautiful and abundant. Amongst them we may enumerate roses, which speedily assimilate to the Maltese type, anemones, violets, hyacinths, geraniums, the vanilla, jessamine, tuberose, heliotrope, oleander, and many others. Strawberries, figs, peaches, pomegranates, apricots, grapes, apples, pears, nectarines, plums, melons, and prickly pears, are amongst the fruits, as are also the orange, lemon, and nespoli or Japanese medlar. Vegetables of various kinds are abundant and cheap. The prickly pear which is the product of the Ficus Indicus is largely consumed, and the cactus on which it grows, and of which four varieties are cultivated, might well be largely used for fences, as the *Opuntia Mayclia* is in many parts of Sicily. Only Nice can surpass Malta, which is in truth the modern Garden of the Hesperides, for oranges, and that only so far as one or two varies are concerned. The fig-tree of which several varieties are cultivated in Malta, is justly prized on account of its juicy and abundant fruit. The first fig which is called "baitra ta San Juan," or "St. John's fig" ripens towards the end of June, on the 24th of which month is the feast of St. John. This fig is of large size, and is succeeded by other varieties, which ripen towards the latter end of July. To prevent the premature fall of the fruit, and with the idea of hastening its ripening, the process known as caprification is employed.

A cluster of wild figs is suspended amongst the branches of the cultivated variety by means of a plant *Ammi majus* called on this account Dakra, the wild fig-tree bearing the name of Dokkara. Numerous diptera (*Cynips*) become covered with the pollen, and convey it from one fig to the other.

The oranges of Malta which are largely exported are justly renowned. They are in season from November till April, and of the ten varieties here cultivated, the egg, blood, and mandarin are the most highly esteemed. The latter is a favourite with the Mandarins of the Celestial Empire; hence its name. Oranges, lemons, and other fruits to the amount of 100 tons are annually exported. Two specimens of each variety of the orange, whether imported or native, are carefully tended in the Botanic Garden, and cuttings may be had on application. Grapes are cheap. They speedily become watery, perhaps sowing to unskilful cultivation, with the exception of the delicious Pine-Apple and Finger varieties, which thrive well at St. Antonio.

During the government of Sir H. C. Ponsonby, the tobacco plant was successfully introduced, but an attempt to naturalize the cochineal insect proved a failure, for want probably of the proper species of *Opuntia* for the support of the insects. Many mulberry trees were destroyed some years ago. Silk of excellent quality can be readily produced, but financial results proving unsatisfactory, the rearing of silk-worms as an industrial occupation has been abandoned.

The best thanks of the auther are justly due to Dr. Gavino Gulia, Dr. Cousin, Dr. Vassallo, Sig. G. A. Pulis, and others who have most kindly given him much valuable assistance.